TOTAL THEORIES OF THINGS AND LIFE

Total Theories of Things and Life: The Infinite Existence of the Universe and Our Finite Experience

Published by Epiphany Globes

To request permission, contact epiphanyglobes@gmail.com.

ISBN: **978-3-00-085000-4** (Paperback)
ISBN: **978-3-00-084999-2** (eBook)

The main text of this book is set in 12-point Garamond.

First Edition October 2025
1 0 9 8 7 6 5 4 3 2 1

Disclaimer

The views, arguments, and interpretations presented in this book are the author's own and do not necessarily reflect the views of any institution, organization, or entity. This book addresses theoretical and potentially controversial topics intended for academic, research, or educational purposes. The author and publisher assume no responsibility for how readers may interpret or act upon the information provided.
This book and the Art cover design **do not** include any AI-generated content.

TOTAL THEORIES OF THINGS AND LIFE

THE INFINITE EXISTENCE OF THE UNIVERSE AND OUR FINITE EXPERIENCE

Pascal Natis

To the world, and one base and foundation to manage and create anything for our future.

Table of Contents

Preface

The most fundamental questions of human evolution remain unanswered. What is the origin of anything beyond simple explanations? How can we live and fill the void of seeking curiosity? What is our past, and what will our future look like while we are still looking for the answers that give us the agony that will never pass? When are we ready to accept our reality —the way it is, with many layers of alterations and behaviors? Why are unlikely hoods still likely? This book is highly theoretical and fictional, but it does include proven phenomena. This, however, does not exclude the plausibility of the named possibilities. This book is based on multiple perceptions, concepts, and different perspectives on humanity and reality to preserve factual and neutral explanations, guided by theories of uncertainties of any kind affecting the Universe, which is the past occurrence of anything that can be perceived in this instant. This creates a foundation that any interpretation might fill a void in understanding anything.

It is up to each of us what we might build with it, or what we will all collectively build together.

This book is based on my thought processes that regard science, humanity, philosophy, and psychology, and in combination serve as a base for further theories to be built upon. It serves as a partial theory of a lot.

Moreover, I have created a suitable playlist that reflects the intricacies of this book and is adjusted to an average reading speed.

The Link:
https://open.spotify.com/playlist/6yYXbv7QDT8IFfUNsBfszg?si=26e78e75176d4696

QR-Code:

An Introduction to the Beginning of the End and Beginning — The Boundaries and Restrictions of Comprehension, Perception, Being, the Past, Present, and Future in a Looped Model

In a relatively big world, there are a relatively large number of reasons and a relatively large number of truths and lies, wrongs and rights that are not perceivable and differentiable by every individual. Our understanding of being (the meaning of life) and of knowing has not evolved sufficiently, or perhaps never will, to comprehend certainties and possibilities. The past provided us with philosophy, logic, evolution, history, and more to this very day, to fill the gigantic void of knowledge until we are prepared for the wave of another dimensional power, an unknown natural phenomenon, to fight for our persistent existence including preserving the knowledge we saw through the small slit of living not past the record of (as of 2025) *122 years* as an individual.

Too short to uncover and know everything about the universe, us, mysteries, and phenomena that might likely occur in a distinct time frame. A legacy that transmits everything we have accomplished to the next generation allows us to preserve our beliefs and knowledge, despite slight to substantial setbacks, encompassing a slight to significant deviation in perception, defined and shaped by disparity and influenced by environmental factors and beliefs.

Suppose beliefs evoke a world and evolve with additional beliefs. In that case, these overlapping beliefs will, in due course, become the perceivable reality of all individuals, a group of individuals, or a single individual. Everything is right and wrong, and in totality, the truth and lies reside within.

In any case, when we look around, reality exists and is forged by time, elements, and us, leading to the current reality we have created, which traps us in it. These are the ingredients to make anything, even the laws of nature, God, or another creature, so intellectually superior that we question the reality we reside within. Something is above us. Believing in something different indicates visions caused by the environment, development, and hence reality, and eventually becomes a belief and truth extracted from environmental circumstances and previous beliefs combined, resulting in preferences, behavior, and actions in a repetitive way.

For that reason, we all pursue a broad range of accessible areas, whether science, facts, religion, further streams of information, or the plain reality we live in, to grasp vast knowledge in a generic way.

A plain reality is the construct of dependencies, choices, and multiple dependencies, declared as a standard, for instance, a washing machine or vacuum cleaner. It describes the banality of reality without slight to severe interventions by the questioning individual. Every individual and living being is affected by reality, which alters perception and choices and adds further changes to a shifting reality that may appear similar in some ways.

Everything contributes to change. Everything we are, we have been, and will ever be, is an explanation so incomprehensible that not even one person will understand it until the end of time.

We would rather choose a reality filled with what we are already aware of and our limited understanding, because we perceive simplicities which we recognize as difficulties, due to cognitive restrictions and environmental influences, as well as suppression.

Yet, here we are still craving a clearer understanding, and thus, referring to the previously stated point, a cycle of craving and ignoring. Humanity is a large network of inhabitants with tinier intertwined networks for communication and exchange, inhabiting a spherical base, acting as a propellant for limited perpetual support in human evolution while advancing or reducing in incremental steps, resulting in an encoded shift of reality, but not a filled pool, and rather a preference for choices of connections and actions. From a biological standpoint, we are unable to see beyond something more profound.

Nothing is right or wrong; it is a property of reality and time — a formed structure defines the future structure. And the structure consists of our beliefs and proofs. That is why "thinking outside of the box" is quite limited, due to the unchanged evolution.

There are no rules in nature, and that is fascinating, except for those obeying processes that create biological life and the facts as we know them, or, as previously stated, belief. It can be described as picking a grain of sand out of a pile that is proportional in size to the current Universe, meaning if something is astronomically larger than the grain of sand, what about the others?

Answers in a pile that infinitely grows and keeps growing, while we stop growing in comparison. To perceive something means that, regardless of its origin, the perceivable physical existence is the proof of processes in a repetitive scheme to incrementally advance or sustain the physicality of the named existence, or to decide the point of creation. This is why they are so limited: space is a vast spectrum of different, minor to major, concise laws, creating a chain of logic we can grasp.

Absolutely nothing is impossible; otherwise, we would be the only living beings. We are limited by ourselves and by the resources on this planet. It will take time until we discover different phenomena and resources in outer space.

This exact reality is marked by survival, due to our biological instincts. It is marked by us and the laws of nature, a power created by the laws of another nature above, a Mother Nature. We could live until we step beyond the border of limited understanding and serve a higher purpose: to comprehend a possible *limited truth*, eventually the purpose of choices, and the freedom to be anything and to find meaning. Any form, any power, any force, and any element to forge a new reality.

The time we live in had a beginning. The question is, if its existence is repetitive and looped. A star vanishes into a black hole, and the black hole itself ceases to exist. Maybe the Universe itself can be considered as a star.

Perhaps we are living in an infinite, finite closed loop, a universe collapsing within itself to neutralize the unbalanced state, like a rubber band at its maximum stretching point before tearing and snapping back and stretching in the opposite trajectory. It would essentially create a similar, repetitive state, with a slight increase in length repeated ad infinitum.
The trajectory is not of importance, but rather the previous and following shape. They would match.

The Universe would behave in a similar state by inverting itself.
If no cardinal directions are defined, it would proceed with the same results as the previous inversion. Out. It is fun to think about when it could be even considered an inversion. Gravity, a strange and looped one. Loop Quantum Gravity and the Universe is a bouncing house.

Einstein's theory of general relativity predicted that once an object falls inside an event horizon, it ends up at the center of the *black hole.* At this singularity, it is completely crushed. At this point of singularity, gravitational attraction is infinite, and all known laws of physics break down, including Einstein's theory of relativity.

It is as if the laws of nature do not always obey themselves, causing discrepancies that affect the general order of other laws —light, gravity, and time — through gravitational time dilation. So, if the universe is perfect —its current state in our finite existence — it becomes imperfect, a significantly altered state at a point where we do not exist anymore, as if the Universe is unaware of anything other than forces and disobeys itself. It might be a delayed explanation for the reincarnation belief, in terms of new organic lives.

Everything that will happen has happened, and everything that has happened will happen. A loop. That is a plausible explanation for why we have no certainty of previous occurrences before the Big Bang. It would mean that the end of time, perhaps the collapse, is the inverted process of space, time, and reality, inverting itself in the opposite direction as an alternating process.

The lowest point, or halfway between inversion and extrusion, in its collapse, is a flat plane, for instance, leaving the singularity. It might be a phenomenon in which everything defining previous, current, and future laws is compressed into a plane.

A continuous and infinite collapse is also a simultaneous creation. At the same time, the theoretical "Big Crunch" is the explanation of an instantaneous inverted process of expansion occurring at an uncertain time and progression, depending on extrusion, inversion and the gravitational forces, only that in theory it would create the previously described looped state and thus create the same previous, present and future occurrences such as the Big Bang, celestial objects and organic life only with a varying amplitude.

It is unclear whether it is a perpetual and infinite process. The existence of the Universe appears infinite to us, but is rather finite over a long time span.

So, what is the *actual* infinity? Is it the infinite and eventually finite Universe? A possibility might be ignoring the laws of physics disproving a disproven perpetual principle, the laws of thermodynamics, that eventually, after the inversion to the final singularity, in which the inverted gravitational collapsing force is not strong enough to extrude with the singularity in the opposite direction, all processes of the singularity cease. The singularity collapses into itself and vanishes —or maybe not.

If the processes alternate in time proportional to their lengths, it could indicate that the Big Bang responsible for our existence is perfectly balanced to create our organic lives on planet Earth. And after Billions of years, a collapse and inversion, a collapse and another inversion until the universe ceases to exist. Or perhaps at the last possible moment, at the collapse of the last singularity, or every bare singularity, the gravitational force cancels out the inversion, causing it to vanish, sustain, or propel it in the opposite inverted trajectory due to an enormous amount of energy, resulting in a plane that would be pulled and pushed in one direction once again and start the process like Newton's cradle and the conservation of momentum and energy affected by gravity, an important example in physics.

The complications of thermodynamics, however, make it impossible to simulate the infinite-finite conversion on most planets, including Earth.

Something before the initial singularity, as small as a grain of sand packed with an unimaginable dense amount of energy, unstable and expanding. It can be considered a newborn universe because of an interruption of the full collapse, which would otherwise result in being dispelled into nothingness.

Expansion again, breaking the laws of thermodynamics for a brief time frame, which temporarily wouldn't exist in the majorly ceased space-time or singularity, because not even the theory of relativity is a defining factor anymore, potentially causing perpetual existence of the Universe. It might be the reason for another, but rather a varied Big Bang, at first. So, in what looped state do we exist?

A law that could only be changed at the last possible moment.

A rubber band. Instant collapse and inversion, a stable state —the singularity, the collapse of the singularity, and the initial force causing initial expansion to the peak of expansion, later ending in collapse. Perhaps even more.

The existence of the universe can be compared to a child on a swing, except the child can swing forever by coming to a stop a few times in the middle, after every swing. It slows, stops, takes a break to replenish energy if it is metabolized, then pushes forward or backward (depending on its last state), slowly gaining momentum and space until it stops again halfway through. But each time the swing pulses get stronger until they reach their maximum amplitude. The maximum amplitude barely touched the beginning of the amplitude, resulting in an inversion. I would consider this theoretical phenomenon as a growing and decreasing state of the singularity. The question is: where is this energy coming from? An accumulation of defied laws?

It might be a completely lossless recycling process, potentially the inversion always gains the same amount of energy to escape the collapse of the singularity, just by pausing while the forces encounter. The universe's size is limited in an infinite room. What could that room be? Is it a sphere? Is there actually an infinite space outside the Universe?

Thinking about it, we are encapsulated in different systems. Multiverses in a tesseract, the Universe we live in, the galaxy, and the solar system. But what happens *really* after the collapse? Is it an infinite growth of space after every collapse? Will it collapse because it reached a physical limit? What if the inversion slows it down, and the Big Bang is the cause of a limited expansion? What if, at an uncertain point, the laws of physics become incomprehensible, even for itself, so the singularity enters a full collapse, as mentioned before, and vanishes into nothingness. But what if that nothingness is another space?

Essentially, the singularity becomes severely inverted so that it doesn't exist, but at the same time tears space and time, essentially starting the process of "the first" singularity once again, if it is looped. But from a broad range of theories, there is a harsh one. The Universe would cease to exist. But why should something vanish in its entirety without leaving something behind? No conversion, no message, no new reshaped version, just blank nothingness. Then what is infinity, if it doesn't exist per se?

Nothing can be perceived as infinite, and nothing can be achieved infinitely. Nothing and no one can count to infinity; infinity is not a value nor a number. It is an unproven belief. We, as humans, can make up or discover concepts, but not always in a chronological or specific, categorized order. Is it hope?

The harsh truth is that our existence is finite, no matter how we evolve, because the Universe — or the sun — would end everything and hope for something else to persist instead. Then what is the concept of finite? Is the Universe perhaps *egoistic*?

Everything within itself is a toy for amusement. Many concepts in outer space are reflected within us. Essentially, the Universe plays with various things in a strategic manner, and we strategically play with toys as well, and that definitive way is boredom or curiosity. Our biological instincts hinder us from behaving neutrally or from obeying strategic orders, such as: "Play and interact with A, then proceed with B." Well, the Universe created its own persistent laws, and the belief in discovering anything through science is proven, but not creation; rather, discovery through interpretation.

We have only scratched the surface of this vast phenomenon, the first layer of many.
Has one ever noticed a familiar aspect? Order, belief, belief, order. There is a noticeable difference in behavior. The Universe obeys, and we obey because we believe. The Universe doesn't obey in its collapse. Such as deciding to survive, an instinct to disobey, and the exact same instinct we all have.

The Universe is thriving in chaos, and we consider it order and law. So, what is this collapse? Is it sentience, emotions, and the proof that God is the Universe? Then why should the outcome of a full collapse result in perpetual existence? Because it can. Because it wants. Disobedience is the result of a decision—the decision of living. Theories about entanglement and Quantum Physics might be the key to understanding a small part of it. We partially understand it, because it exists. It is a gate that opens a new realm from traditional, intellectually prohibited understanding. We can understand only so much as we have intellectual capacity, creativity, and a broad curiosity underlined with a repetitive question: *"Why not?"*. Even though our different perspectives fulfill a portion of our meaning of existence, there is no fundamentally universal answer, so that everything can be based on that very thought. A thought is unstable if shared. It introduces chaos.

Realms of Reality and their Infinite Properties in the Universe

But what about the reality? The veracious premise of branches, akin to that of a tree, can be applied to the reality we can perceive. For instance, a sphere represents reality, while inside that definitive sphere lies a smaller one. A branch extends from the outer shell of the smaller sphere at any space and time point. These branches are proportional to the gravity of the inner sphere and are, for that reason, eventually curved but not spherical. These branches could overlap in different planes and miss, or connect to their point of origin, or the opposite. The length of the branches determines the time to completion. The branches could potentially describe our possibilities and somewhat seem infinite, due to the uncertain number of them as well as the duration of each, hence infinitely looped. If the branches connect and intertwine, however, it is a closed thought and is reflected as a tiny message in natural phenomena and laws, conceivably like an encoded message and order, and another example of mirrored laws of nature in organic life and us. Principles of the universe are reflected in our behavior. Access to the amount could fluctuate, depending on cognitive capacity and environmental factors. This whole sphere could be perceived as a fracture of our reality and the laws outside, above, and within it.

What, if *anything*, is reality? What if everything we think of is real, but not necessarily in this time stamp or reality? If something is beyond the theoretical multiverse, is it another universe, a tesseract, or 4-dimensional behavior? Are we in a linear constellation of multiple universes, behaving similarly? Is it a capsule containing everything, and the capsule is floating in another galaxy? Perhaps many universes in the multiverse are arranged as the outlines of a tesseract, which is in another dimension – a loop that continues.

I am aware that it is really complicated to determine what the nature above us looks like as a fractal simulation or in any other description, due to our differences; a uniting explanation is unlikely. There is an uncertain finite abundance of variations, and a small portion of it falls into the spectrum of discoveries, determined by time.

We only search for a meticulous answer, and if one is found, it controls all processes related to life – including laws, boundaries, general behavior, and emotions. Are we in a locked infinite loop, but only finite for a brief time to become duplicates and restore infinity? If we theorize various aspects, instead of believing and hoping they will become reality, nothing will happen. Why are movies regarding heroes and other beings from outer space? These movies and concepts wouldn't probably exist if they were reality; instead, they are a reincarnation of history in our present timeline. We don't create something if it already exists in reality. It is something we hope for —the hope that keeps propelling individuals to strive for achievable goals within this limited world and reality, while finding meaning in it. It is only as restricted as one wants it to be.

We substitute our perceptions with restrictions. Deep in us, we know that we are alone in this vast galaxy for now. There are only billions in this enormous space of light and darkness. Why can't we reach the physical edge of space without proving it? Because what would happen if we did? *Anything* that appears real to us would break down.

It is understandable to wonder whether we live in a simulation. The operators would have a different reality or exist beyond this galaxy.
Are they simulated too? If so, we would all live in a closed loop, even the universe. It doesn't matter if we are simulated or not; the fact is, we cannot change what we have been, are, and will be, except for the future, which lies in our hands.

We might be living in a closed loop that has different amplitudes and corresponding particle arrangements. Is the answer to a multiverse a misaligned and looped universe that duplicate itself outside of the singularity? Is this duplication projected in our thoughts and beings? Is everything infinite? Is there an accurate volume estimation of space now? Why would *nothing* be a plausible answer? If infinity and size are irrelevant, what would be left? It might be nothing, it might be anything, it might be the infinity of nothingness, any answer. We can never understand everything; only to a limited extent, and never the exact amount that might exist in the finite patterns. If a lot is unknown, then anything beyond ethical definitions is right and wrong. It is only set by our preference as beings.

We break a possibility by assumption and create a new one from it. Can any theory be true and wrong at the same time? Is everything similar to Qubits in superposition? Wouldn't that be possible with the universe, too? Right, Qubits exist in this reality and dimension. They might be proof of superposition to natural phenomena, opposites at the same time, torn apart and reassembled. Yes and no, right and wrong, here and gone, dark and bright, forward and backward, left and right, up and down, far and near. Everything has opposites at the same time in this universal reality; otherwise, the laws of nature cause a collapse. The reason is due to instability without them. It leaves us a choice. But the universe decides by its laws. Opposites create neutrality and cancel out their directional path after an unknown time. If there is an equal amount of right, then there is an equal amount of wrong. Superposition exists everywhere and can be considered as neutrality by itself.

But what lies inside the superposition, except 0 and 1? What is the barrier dividing them? The middle of the superposition could be a crossover between 0 and 1, while being considered as the neutral zone, more or less the decisive point, due to rapid shifting, which results in equalization, because the choice can be perceived in both states at the same time. So, the zone might use a specific pulse width as well as a controlled modulation that releases with the tendency towards one choice. The environment influences the state.

Even though everything is in superposition, superposition affects itself. It is similar to the origin of the Universe; it only affects itself. And for us, a conscious experience arises whenever a quantum superposition forms, which is continuous.

Regarding behavior as a superposition, it is only displaced and dispersed. That is why an alignment of right and wrong for us as humans will never succeed in being deployed. The state is continuously altered and combined in a different time and space relation. The corresponding opposites are not unifying and therefore create different approaches, opinions, and wars. The same side is chosen. It is a natural phenomenon, reflected in us humans. Forming two states. A big positive charge and a big negative charge. It creates an explosion, lightning so gigantic that it ruptures and destroys everything.

The war between two opposites is equalized by the physical encounter, temporarily balancing and, again, restarting the process, ignoring the uneven number of neutrons and the superposition for a brief period. Quantum entanglement may also play a significant part in this constellation. But the equalization is somewhat similar to a pattern. It appears to incorporate human aspects for creating a temporary image that we define as life and reality, yet the visible degree of this temporary picture differs from others. At an uncertain time, order tends to reverse or change, often in a chaotic, entropy-like manner. A usual explanation is the altered behavior due to the environmental factors, which include behavior of any kind as well, e.g., particles and forces.

It is more than that. Entropy is a literal *universal* phenomenon, and it is reflected in our behavior. This exchange advances each time and becomes stronger. The question is… when is it immensely unbalanced or so intense that it separates both charges and disperses them out of reality, as an inversion? Will that definitive charge be sufficiently strong to invert reality and be the reason for that looped process? Innovation and evolution are not even comparable to what lies beyond this world, this galaxy, this Universe. We create stories that can be as big as their creator's imagination. The truth, however, is shorter but also proportional in size to the lie. Wrong and right, truth and lie, the opposites that dictate us. Either we become immortal, or we will be extinct. Nature is the ruler of Earth. The immortal cell chooses immortality. Nature gets to decide who will have a higher potential and who won't. The behavior of individuals affecting others is an example of nature. Everything in this world can be considered nature and natural.

The elements fought for persistent existence. And millions of years later, they still exist, symbolizing a formed, stable state with this planet: a symbiosis and a calibration. We use the inorganic ones, but we don't rule the world; we are being ruled by the organic selves and nature. Perhaps the initial God is the one who created us, or the ancient nature created God, or nature created nature and God, or galactic nature created our nature. Beyond this universe, laws that are familiar to us don't exist, but rather as an ongoing concept of a conceptual reality. They might be inverted or the parallel act of a looped creation.

The outer laws perhaps create the following form of this universe and a parallel reality that is not aligned with continuous time and is bent by gravity: the gravitational time dilation and the relativity. Perhaps the same law exists beyond this universe, and conceivably, it could be the opposite and accelerate time. Time is the ruler for the laws of this universe, dictating them, but also making them controllable.

A constant might cause deviations, cosmic rays might cause mutations, and eventually, different living beings. Our divergence on this planet. No right and wrong, but the nothingness and existence created by it. A powerful being or nature. We will never know, but one considered God —or the creator —to be still alive on earth. Mother Nature. Every act, every person is a fundamental, critical piece, cooperating to create something. Why do we have an urge to know and evolve? Isn't it clear? Do we need to fight and protect ourselves against something or someone? Possibly… against ourselves.

Is the path of modern technology and society destroying this planet and Nature, for the next generation, because we are on the wrong planned path and need to *pay* for it with our lives? Even though nothing is inherently right or wrong, what is this posing? A preference of nature? Is the orbital dance between right and wrong causing a neutrality in which natural laws could bend reality and beliefs, and cancel out right and wrong temporarily? Was it an unforeseen mistake? Is it a planned climax to cancel out the possibility of such a path *ever* again? Is a social and geological collapse a rebuilding mechanism while living in another mechanism?

A mechanism is a law defined by its creator. The universe is a law, hence a mechanism. The Universe is a creator, thus a creator. Our devices are mechanical and obey the laws of nature and the universe's set laws, or as we know it, science and us.

Fact is, whatever happens is predestined, because it could never have happened differently. Even the plain thought of change is unchangeable. There are various individuals who believe in the simulation theory. Well, if we were to live in it, we wouldn't know it, because it possibly conflicts with one of the reflective laws or general behavior. If one is *something*, it cannot be explained, because that state is currently in use. Discovering the very self is a continuous process that is constantly altered by the environment. Life is a session.

If a caterpillar becomes a butterfly, it could be aware of its previous state but not of its ongoing or final one. This analogy can be applied to knowledge as well. The continuous present is mandatory for the

future. But looking at the past, the previous present suddenly appears reachable and explicit. It is somehow explainable.
This is the very concept of developing and becoming. No matter if an object, a life form, or matter itself, it depends on time to become whatever state the laws define at a specific point in time.

In summary, everything is palpable and explainable from the past after a certain amount of time. A hybrid combination of time and emotions gives us the ability to comprehend past occurrences. But not the future present. This is why it can be considered as predestined, because there is no fixed outcome – only the predictable one. If something is predictable, this fixed prediction causes all other factors to change, and the future deviates from the previous prediction. They appear unfamiliar when they occur because they have been ignored.

Absolutely anything could be predictable if we could defy the Universe. It would essentially mean we surpass all laws that exist and will exist. What happens? Exactly, everything would vanish in an instant. A fixed state – everything is finite. There cannot exist something larger within the only space. Another thought-provoking concept is infinity. There is no way to prove infinity because of the very concept. Infinity cannot be proven by finite processes; therefore, we believe in the concept that this reality is infinite. But nothing can be infinite if infinity is the box we exist within. Because of its gigantic space, it cannot be reached, and we consider it infinite. A space cannot be infinite if it eventually collapses. And if everything ceases to exist, it will always be considered as finite. But something that cannot be seen, felt, or perceived in any way, entirely removes the fact of infinity. Because whatever can be perceived, will cease to exist. The Universe itself will cease to exist, and what happens after is smaller than anything that could be perceived. An outcome is unpredictable within most foundational parameters.

So, the conditions of being aware remove the possibility, but it does not apply if something is infinite. The question is whether we can consider infinity as having a break. A continuous infinity is unlikely. Whatever we do, we can only mitigate it. And if something unstable appears as uncontrollable, the control lies in accessing the base in a physical manner.

Reality and Its Creation of Cognitive Faith and Belief

Once in a while, one might look up to see the sky. They will see it as bright, golden, and dark, with different shades and colors: two options and a shifting neutrality in a crossover. Any condition of the sky alters the questioning of a living being. One might explore their faith further, while the other craves an understanding. 600 BCE on warm summer nights, where the early birth of cosmology began, physiologoi ("men who discoursed about nature") noticed that while most stars appeared to be stationary, a few "wandered" across the sky over a specific period of time.

This is an example of observing an unknown state. With imagination, there were various interpretations of some behaviors of stars and celestial bodies. Believing in a coherent and natural explanation, mathematics was used as a tool for understanding. Whether the interpretations differ slightly or significantly, it is rather difficult to differentiate between belief and "actuality", since any such distinction would run into the same issue. Any individual who was part of observing these unknown states came up with different solutions or explanations based on a theory and, hence, a *belief*. The belief would turn into something factual for a convincing gravitas and be categorized as "true". Mathematics might be a tool to interpret various phenomena.

The foundational concept of mathematics emerged from keeping track of quantities, such as items or physical forms. Mathematics might have started as a discovery —a need to keep track of quantities — and was later built on curiosity, belief, and assumptions, altering the concept of a particular aspect until the answer became plausible. The early stages were a purpose of quantification. The interpretations differed, and numbers or signs were created to assign a single quantity, or, in the case of more than one, an amount. Two separate but identical items were combined into a single quantity or sum. For further discoveries, assumptions and beliefs became part of the mathematical system. The main goal was to convey the surroundings in short, logical explanations. Numbers are far more efficient compared to words, but explaining the thought behind a system of numbers requires sentences. It isn't easy to imagine being part of the explanation that two identical objects would be the sum of two. But that was the definitive starting point of mathematical logic—a system for understanding the perceivable surroundings and conveying their properties and behaviors.

Proof derives from assumptions based on perception and cognitive processing, including other factors that alter them. Cognitive theory posits that one might tend to perceive and interpret situations in characteristic ways that color their feelings and shape their behavior, thereby changing their thought processes and, in turn, logic. How can something be comprehensible if it is not a unifying aspect but rather a separate logic created within mathematics?

Somehow, we see standardized mathematical concepts being disproven or proven continuously, yet new concepts develop with the need for innovative solutions to self-created issues. An idea will be altered until it matches a clear previous "proven" metric and will be refined further until the proof is plausible. We harness the environment's properties and translate them with emotions and interpretive reasoning to convey a "simple" solution.

It is somewhat amusing that our current understanding was conceived out of need, hope, belief, interpretation, and conviction. It also appears to be the reason why analogies are only as limited as the mind. Even logic follows the very same rule. Individuals are frustrated by not being able to prove their concepts with the current advancements in any area that involves determined logic. In a way, there aren't any groundbreaking new methods, such as the Pythagorean Theorem, due to the vast number of problems created by foundational concepts and gaps in mathematical interpretation. These gaps hinder any new discoveries and interpretations.

If mathematical logic is flawless, how come it cannot be reused and applied as a *universal* method? In the very beginning, every theory and calculation was carried out with the thought of understanding. Humanity could only achieve that by focusing on a specific aspect and interpreting it with previous methods. After iterations of trial and error, the result may come close to the "original" inspiration.

Within the Universe, there isn't a single "perfect" circle or sphere. That depends on how *"perfect"* is defined by any observer and participant. The only way to utilize the shape's properties was to create a universal mechanism to maintain the previously defined properties across different sizes —Pi, the concept of scale.

No celestial object is as round as any other. The object has impurities, dents, and various patterns. Even incorporating these aspects, we use only the readily apparent property, namely, the circular or spherical shape.

Countless individuals have dedicated their entire lives to defining this concept of shape and creating a general shape recognized as "perfect" and correct—the circle and the sphere. These creations are then converted into formulas to reduce the visual comprehension and instead specify them in text and numbers. The logical gaps that follow from these definitions and interpretations stem from the different ways a concept is perceived and translated into action.

Only the creator, successor, and the willing observer(s) agree on the basic foundations of the creators' concept(s) and recognize the idea as accurate and a new standard. It is the power of the majority that is accompanied by persuasion and thus creates a foundational belief system that is "correct".

The way is still being interpreted by previous definitions. It gives the impression that mathematics has reached its *final* state of significant evolutionary progress.

A Component of Another Consciousness

We only have a finite number of possible thought patterns and combinations. Time is finite. Our time within the time is finite. The dispersed energy in this universe is finite. Stimuli are finite. The particle combination is finite, and thus repetitive patterns appear, which will, in the next concept of mine, represent the intercept and overlap, which are marked by similar repetitive processes. But the size can at least give one the ability to explore a small part of it.

To explain thought-provoking concepts, one might use familiar ones for a coherent understanding. The mere fact that humanity replicates phenomena in an interpretive manner and has struggled to evolve continuously might raise the thought of an individual moving within a limited and uncertain radius, which is fixed and cannot be altered by any external measure, due to the external alterations and perceived reality being already incorporated within the radius. The diameter is the finite time, and within these radii lies a map that represents a headline for every moment in time, whether a description or an action of any kind. The headlines and actions are accompanied by distinct patterns defined by time and space. These patterns are not definable, since they represent something different in any physical matter that is affected by the circles involved. This concept also becomes familiar in a collective manner. If we use our current representation of circles, a collective pattern emerges, and we will have multiple intersecting circles, many of which partially experience friction due to the intercepts.

What happens when the collective circles intersect is quite fascinating. The partial or complete intercept creates an exchange that, for a certain amount of time, forms a new geometric structure. This is the *friction* of circle intercepts. Some might appear linked by a partial or complete intercept. But how would each circle that represents an individual have any correlation with the Universe? Well, friction is also defined by gravity, because each circle rotates along a different combination of the X, Y, and Z axes, while time can be considered an axis itself. Each circle has a core that is strongly positively or negatively charged, affecting circles within the radius of the whole scope.

The gravity acts in a circular pattern. The rotations might also match some patterns within the other circle(s), creating a unifying aspect. This concept combines any physical state, such as brain activity, and, by extension, thoughts.

Everything is physical, even if it does not appear that way. I consider this law as: "Fixed Patterns of all Kinds of Behavior Reflected and Absorbed by All Kinds of Particles with Influenced Forces in a Finite Time/Space Relation Dictated by Particle Patterns" or the fixed recurring unpredictability dilemma.

Assuming an uncertain moment in the loop of the Universe (growing and decreasing) results in the same pattern we live in, or a similar one, practically everything that happened will happen, and everything that will happen has happened. Everything that happens has happened, and will happen, if the Universe were a loop. This law even disregards the "Reality Transurfing" theory of a past Russian Quantum Physicist. Vadim Zeland's Transurfing theory holds that infinite versions of reality already exist, and by shifting one's inner state, thoughts, and intentions, one can "surf" to the preferred life track rather than forcing change externally, which would create resistance. But in this context, the change is relatively constant and cannot be influenced if it is a set particle combination recurring after a specific combination, which means altering the change is not a change but an insight into a perception.

A Brief Summary of Typical Human/Primal Behavior

Without considering the previous chapters, explaining how we work and how we work collectively in this world is sometimes a mystery we attempt to solve through belief and proof. So how do we explain the vast Universe and us? Something had to create the Big Bang. There's no possibility that something conceived could come into existence and presence without an *origin*, whereas it could have always existed. Therefore, religion is a belief and a practice of many groups and religions. Some hold on to scientific proofs. Diversity is a crucial variable in the differences that still create the environment and reality. Humans always create sides to choose from and fight over, rather than unifying and enjoying the limited time we have.

And as Stephen Hawking once said: "We are just an advanced breed of monkeys on a minor planet of a very average star."

Of course we are. We belong to the kingdom Animalia, the phylum Chordata (or subphylum Vertebrata of Chordata), the class Mammalia (and thus mammals), *specifically* the order Primates and the family Hominidae. And in short, we humans are animals with backbones, classified as mammals and, more specifically, as primates within the great ape family, which explains both our biological traits and our evolutionary kinship with other apes. It means most of what we do is a Primal instinct, caused by biologically controlled chemical changes in the brain that cannot be controlled, hence the Primal instinct. We will never know what happened so that it could happen. It was the beginning of space and time. We only know what happened to a limited degree, and we should use what we possess, what we can create and use, because we are not gods in the sense we might imagine. We must make ourselves believe and question; it is something we will never be able to change, but we are destined to follow. We can only create as much as we can conquer, and there are limits. But the limits are unseen boundaries. They are boundaries we made, not ones that are present or will ever be. There are biological, and hence evolutionary, boundaries, but we won't reach them, because there is no way to determine whether we have reached them or will reach them. We always had the urge to bypass our limitations with the help of technology. What we don't realize is that everything we see is built by us, and it all began with minerals, earth, fire, water, and wind.

So, in some way, we harnessed our potential to do something great, greater than any species could have. The question is, when can we keep up with it, especially if there is something that surpasses almost everyone? But there are no boundaries, and that is fascinating. The boundaries are our routines. Many individuals exhibit depressive behavior and typical aspects defined by rank, power, and pre-cut templates on how to live a life, which keep us from reaching an unstoppable state, in which we could achieve anything, limited only by resources, intellect, creativity, or both. We have achieved a lot already, assuming there isn't much more, except *there is*. Of course, we will always reach new heights and new realms beyond comprehension. A lot of efforts reach a climactic point of no return that have been so insecurely reached—Artificial Intelligence sets a great example. Tied to revenue, not security, and not an actual aid to discover more than the human mind ever knew, but instead supplies it with a distraction of high toxicity.

We can't comprehend our conquests or forget them, but we will subconsciously know that they are present. Our flaws are the perception of inbound future threats and challenges, the ability to reason, imagination, and the main components that secure a chain of logical thought and complete comprehension. We are logical and human. Both together result in *irrational* rational choices influenced by previous experiences.

That is the aspect of miscommunication and misunderstanding—we are not comparable except for the biological part, more or less. Every one of us evolves differently. Multiple factors play a significant role in our becoming, for instance, genetic history, childhood, environment, and social exchanges (such as with parents, friends, and the social circle). Nevertheless, it results in a fragmented image constructed from a unique angle in a unique spectrum. No one has a complete picture from every possible angle. Curiosity may fill the majority, but evolving provides a proper understanding of our own being. Comparing ourselves to one another is quite useless unless we fill in undiscovered fragments. Because if we compare ourselves, it always ends in chaos and power plays, and the primal behavior reappears. The powers within the Universe fight as well, but they never vanish, neither do we. We have to be better than we were supposed to be. Nothing turns out the way it was conceptualized by nature or us, even the 3 Body Problem.

We substitute reality with experiences gained from a different perspective, resulting in selecting sides and uncontrollable chaos—the "Homo Sapiens alignment dilemma". The best way to describe humanity is through its *uniqueness*. Otherwise, we are not better than any other mammal or any natural phenomenon. We are the result of a combination of chaos and particles that turned out to be studying themselves. We are the reincarnation of a conscious Universe. Nothing is lost, even energy. It is converted. We can convert every one of our own aspects as well and create parallels to original phenomena; for instance, the bare concept of audible sound from anything, and with it, influence the way we behave and feel.

We receive something pure, a constant shifting pattern within a fixed space, but the fixed space is as vast as anything can be.

The question is: what behavior and patterns lead to a fulfilling life? Time is changing within our patterns.

We can only see a small part of many parallels and don't have a clear view, which results in uncertainty. But we focused on the solved aspects and imagined new uncertainties, which in actuality are certain, even though one might not be aware of it. When something is unclear, it gives an individual a mission. That mission becomes a journey. We are imprisoned between being highly developed and still similar to our biological neighbors and predecessors. There is not even movement in between, which is a new zone.

We invented art, poetry, philosophy, and science, aiming to break through the barrier that we should not pass. The barrier of understanding everything and collaborating to avoid chaos, for which those who are not responsible must be held guilty. The theory of everything poses the risk of losing control.

It was only a matter of time before our early ancestors would develop enough cognitive capacity to find a way to compensate for the truth of untold reasons, the sense of developing thought, and the creation of beliefs and opinions—the start of a trapped plus looped reality that we are still experiencing to this very day.

Why We Follow the Heart

The way we follow a decision or a choice is through the majority and emotions. It is essential to note that we all use something similar… logic. Whether we use a machine that is biased or math, it is pure logic derived from hallucination and imagination. The way individuals decide is based on emotional and experiential past occurrences, which gives rise to the saying "follow your heart".

Our brain has two hemispheres: the logical and the creative. But why do we separate these terms? What if creativity is an advanced process of logic? A jittery logic. It is unique and sometimes indescribable. Philosophy isn't defined as pure logic; it is something profound, something truly unique. If we miss a gap in evidence, experience, and knowledge, we can't — and don't want to — decide, which is why some are attracted to the concept of the majority, which represents the big "trustworthy" heart. Indeed, a heart is powerful, red, and soothing, but if a social and special system is built on it, then another option is not plausible, except for a few. Sometimes, there is a body that is observing the acts of the individuals in the heart as blood cells, soothed and fulfilled. The small percentage not in that heart are bodies — bodies with the generic human anatomy: a heart, a brain, and limbs. They move freely; they wander with their minds. These bodies can never be pure blood cells, except for the fact that they have their own.

They want to *show* the big heart and the big blood cells, that the big heart needs a big body and organs, so the blood cells have a purpose, an elevated level of existence. But the blood cells feel comfortable; it's their choice. The bodies are left to fend for themselves, while only the heart gives warmth. The smaller bodies *dream* about giving the big heart and blood cells a big body. Instead, choose to provide each blood cell with a body. Now all bodies are freely moving and observe the heart. Meanwhile, a key is lying on a table. This key can unlock almost all doors behind the heart. Only a small number of embodied blood cells become curious and fight over the key. One gets it by fighting the other's embodied blood cells, infringing on humaneness.

Now that the embodied blood cell has access to six out of seven doors. It unlocks them. Some rooms are empty, and most have objects. In one room, there is a chair, a table, a book, and an *apple*. The blood cell is not aware of what it should do and leaves, but it is a gift of solitude that this blood cell does not care to live in and enjoy, even for 30 minutes. Another blood cell enters the empty room, reads the book, eats the apple, and enjoys the sunlight while sitting on a simple wooden chair. It wants to know more about the different compositions that make up everything it was not previously aware of. So, it unlocks the big extra door behind the heart and goes for a walk. Mountain after mountain, tree after tree, river after river, eclipse after eclipse, it finds a moss-covered rock that was under a waterfall.

Amazed by the atmosphere, it grabs the rock and turns it over. A strangely shaped key with the embodied blood cells embedded sparkles and is black at the same time. It runs back to the big heart and opens the door into an ample space of nothingness, while the other, embodied blood cells, sleep.

As the blood cell unlocks the second door behind the last, it is stunned by what it sees. A library, a garden, plants, fruit, food, paper, and its family that it has always *dreamed* about. Now it can enjoy all the time, learn the answers to all the questions it's been eager to learn, and spend time with its family. The next day, the doors are gone, and the other blood cells that are embodied think it was a dream and climb back into the heart. The other embodied blood cell, as someone in the other world, though, cuddles with the other hearts, the same size as their own.

Why am I telling this boring story? It reflects our society. Most simple tasks, such as living how one wants, are rather complex for most because of the environment and the obstacles that unnecessarily arise. Therefore, individuals with another perspective, not included in the majority, are constantly suppressed and belittled. They need to hear the dilemma and suffering of the ones who have chosen to live their lives the way the majority does. "When you grow up, you'll see …"; "You think everything is so simple…"; "I don't think so, wait until you get into this… well. Lashing out and suppressing others seems to be a habitable way of living nowadays, primarily driven by insecurity.

This unstable state is marked by insecurity about others, the fear that there is something better. We need to keep in mind that we are just primates in a suit built from experience and evolution, and the wonderful nature we can enjoy, harnessing resources from planet Earth. We exclude untraditional things, leaving corrupted and contagious choices.

The way we live, if no will for change is present, is defined by: Area, wealth, family, intellect, personality, and popularity. Bending this template creates disturbances and distances. The person who suffers the most, or has suffered, while trying to compactly bend this template and incinerate it to leave only a finite amount of particles, which could be the basis for something new and profound, takes a walk to calmly find a different parallel realm to succeed with their concepts and thoughts. This template is not only applied to every aspect of living and evolving. How can we ever grow if there are fixed choices to pursue similar schemes? A scheme creates disturbances and insufficiency for those who are respectfully declining it. For the majority, it creates an organized state that controls various aspects of social and environmental phenomena, while also introducing significant pressure, stress, and unhappiness.

What is, is Not Now

Some might have wondered how to explain the unexplainable. There is a correlation between being something and attempting to describe it. From a broad perspective, using an instance to illustrate the instance results in nothingness. But the instance can be used to perceive a relatively small amount of itself, which our current half-uncertain reality is based on.

A similar concept can be applied to us. We cannot explain something we are, but we can discover the necessary aspects that emerge as relevant. *Who am I?* - the usual answer consists of experience which leads to a final explanation, e.g., "I am a…" This raises questions. How objective are we? This is precisely the very standardized loop that the instance is highlighting.

There is only an aim explanation for something that someone or something is. It implies that, due to continuous development, a living being, as an instance, cannot explain its own instance or being. A usual approach to bypass this phenomenon is to rely on insecurity. Insecurity has a necessary aspect in reflection, with altered perceptions of others caused by the same loop, leading to unforeseen outcomes. Insecurity is one of the many networks that all instances, whether alive or not, depend on. A deceased individual not only affects reality on an atomic level but also on an emotional level, where reality continues to shape in diverse ways. In contrast, this results in alternative versions of the perceivable 'now' and the self.

It is fundamentally relevant because it causes all processes that are unconsciously perceived. It can and will be right or wrong. In theory, this phenomenon is responsible for almost anything, even ethical decisions. Thus far, everything reflects within itself.

No one is in control, but rather the subconscious being embodied in parallel with the conscious, while being highly developed to cause what we call wrong, right, and general compatibility issues among social conventions that create levels and are currently the only "safe" option to realize and interpret reality as it happens, which excludes broad spectrums of unaccounted uniqueness and reason.

A thought on what consciousness reflects in this behavior in every way. Objectively describing what consciousness is raises obstacles to the notions of preference and right and wrong. But perhaps anything perceivable is a fragment of the partially expressed consciousness that is accompanied by primal behaviors and thus *instincts*. Maybe this is why consciousness can never be explained: a coherent explanation does not incorporate aspects ranging from generic to unique and rare. Consciousness might not exist, but it is the interpretation of being and aiming to extend the description of what one is, which, as mentioned before, does revert to the primarily mentioned issue.

No One is Bad

We look at someone and, in an instant, create an imaginary personality. It may succeed in protecting one from something, but it may lead to choices that can impact one's life in a good or bad way, while also igniting a chain reaction of similar phenomena.

Most individuals may be "bad", but understanding the reasons that led them to that point is quite understandable if we take some time. No one is trying to act out unethical actions, but due to circumstances and environmental factors, they unfortunately occur. And the concept of the imaginary creation of personalities is a protection that has endured across our generations. Therefore, some fear and some don't. We don't trust as we used to before a significant change. The point at which it all went wrong and changed our lives forever is untraceable to this day. *War* exists due to leaders who experienced unethical circumstances and want to compensate for those by replicating these actions to neutralize and balance their mental health, and using power, rank, and authority. In childhood, being abused results in utilizing the authority for compensation. It acts like an autonomous mechanism that balances an individual's suffering past and shines a spotlight that was once small and dim.

The problem is, our brains have no perception of how much abuse occurred, but instead categorize it as traumatic, leaving the individual no option but to counter it by doing the same, hoping to equalize it again. Overdoing these actions, such as revenge, will eventually result in psychological problems, like mental illnesses such as depression, and hoping to compensate for it with the same formula as before. This compensation is noticeable in personality, behavior, and actions. Other experiences can lead to different compensation stages as well. The compensation is a fixed protocol in the brain, executing it until an uncertain point in time, while remaining resistant to external stimuli that might trigger an easement.

Individuals always compare themselves. Religion, interest, and more. Persuading one to be pulled down into insecurity, not to feel alone. The fight creates a social exchange, whether harmful or not, and is therefore described as "social". It is something that the instinct is dictating. Belief, which creates most of our perceivable reality, becomes an insecurity because it is neither factual nor proven. This results in a conspicuous, uncontrolled chain reaction of chaos. This insecurity can extend or spread within other groups of individuals. The previous belief will then be altered by the number of individuals in the group to create a temporary statement, concluded as proof with a short definition, whether it is right or wrong. This process is still ongoing to this day, highlighting that neutrality is not common. Not only does this create a chain reaction, but it is also subtle, and it contributed to most changes, slow evolution, laws, and everything we see around us today. This insecurity can also become a threat or a form of manipulation, hence the familiar concept of blackmail.

Nevertheless, the individual exposed to a similar environment that might resemble a stimulus or trigger will not only recall emotions and experiences but also alter their behavior, which is sometimes considered a "unnatural" moment. This also contributes to insecurity, as it affects their surrounding individuals, though it is a worrying insecurity. This is a simple representation of how something small but fairly broad, like insecurity, can lead to multiple and parallel processes that affect anything and anyone and shape the thought patterns.

Dependence on Independence – How a New World Might Solve Everything

We have often enough asked ourselves how we could escape all the chaos and suffering that slowly covers the world in a blanket of lies, egoism, hatred, and instability of all kinds. We suffocate due to our very own nature. When we explain something with all our uniqueness, this blanket suddenly becomes transparent and dissolves in the atmosphere of our exhaled air.

This reality and current constellation of life are not only decreasing the rate of our evolution but also reversing and defying it. The key to solving the majority of our problems is something that has manifested itself for centuries now. What could be the theory for *everything?* We cannot find it because something else is continuously trapping and restricting many specific processes, even though we are not sufficiently supplied with it. It is perceivable in various and different ways, but the concept still remains. It nearly controls all of our aspects and decides whether we survive, suffocate, die, or both. It represents power, influence, and health, but makes any individual, no matter who, prone to various types of sickness and disease. What is that vast key that dictates all of these aspects? It is money. Humanity goes haywire, delusional, and lies only to protect what is fluctuating and somewhat insecure.

This could be a collective tool, yet it goes to show that money has transpired as a tool for power, leadership, or support for a variety of individuals who have a higher quantity of it, whether ethically earned or unethically.

This encompasses the history and reasons behind various outcomes, as well as the controversies and issues currently arising from them, and is based on factual perception and reflection rather than primarily on establishing what is right and wrong. What is humanity's next move? Is it to concentrate on industries and offer goods to consumers who are financially constrained? And, as a result, cannot afford what would undoubtedly give them a clear purpose, allow them to express themselves, or stimulate their minds, and perhaps all of the above and more? Hypothetically speaking, there is a causal link to significant difficulties associated with the said subject. In an unfamiliar context, a solution seems improbable due to our interdependent reliance on this subject.

In short, there has to be an abundance so enormous that its meaning physically vanishes and becomes an independence. This unfamiliarity is likely threatening and frightening, but has undiscovered advantages and fewer disadvantages if incorporated and meticulously planned. Incrementally, making its purpose meaningless for the collective good. Surely, proposing such a thought-provoking idea raises problems, like where to start.

Beginning with health, a money-free world would allow any individual to receive the ideal, transparent treatment necessary to stabilize the patient and potentially extend their lifespan on this planet. This way, analysis and treatment would enable deeper, more transparent, and more coherent studies to implement new methods for safer, more efficient treatment, and the improvement of processes would be significantly advanced. This also applies to medication and general pharmaceuticals.

Surgery has always posed significant risks in certain operations, due to their rarity. But this experience could be expanded and studied further by introducing the possibility of repetitive procedures, which would be made possible by a broader number of patients. Assuming it excludes harmful procedures, such as expensive and detrimental medications and supplies, it would accelerate improvements in future well-being.

These examples are just the dawn of a better future. Why? Disparate studies and treatments for cancer treatment and prevention all share a similar result to this very day. There is no non-invasive, safe, and proven method of *fighting* cancer. Many might presume that cancer is the result of a mutated or many mutated cells, due to toxic gases or other toxic elements, ingested, inhaled, or exposed to. Undeniably, this can occur in some cases. However, the majority of complications to why cancer exist do not arise from unhealthy nourishments but consequently the mere fact of the way we live. The tremendous amount of unhappiness, sadness, stress, or prolonged pressure and hence stress makes any individual prone to the disease in uncertain advancements to an uncertain degree. This could be a child, a young adult, or an elderly individual.

Ultimately, one might ask how this fact is actually the primary reason.
There are *two* foundational states of the autonomic nervous system: the sympathetic nervous system and parasympathetic nervous system state. The sympathetic state prepares the whole body for stress, including increased heart rate, increased blood flow to muscles, and a higher oxygen supply to all organs, except for the sole supply to muscles. It correlates with any form of stress or distress. The parasympathetic state, on the contrary, does the opposite. Heart rate and, thus, blood flow decrease, and the body returns to its "idle" state. It correlates to relaxation.

Going on, suppression, traumatic experiences, and constant stress will promote the body into the sympathetic state. The outcome of the accumulated experience and behavior is the body's inability to recover or heal. Prolonged exposure to or involvement in stress can lead to further complications. Prolonged stress and psychological illnesses reduce the adrenaline level, which is a stress hormone and necessary to provide glucose to the muscles immediately, and is an example of the fight or flight instinct. The constant utilization of adrenaline reduces the adrenaline level. The daily routine of waking up early, along with sleep deprivation, and working or carrying out specific tasks every day, while entering a state of unhappiness, is the result of a prolonged stress state. Many individuals do not realize the danger it poses and consider it generic.

And what is the correlation with cancer? Insulin and adrenaline are mutually dependent to maintain a healthy state. The insulin carries the carbohydrates into the body's cells, where they are stored as glycogen. The normal adrenaline level is responsible for taking the excess glycogen or then glucose after the cells' energy demands have been met. Too high an amount would destroy the cell, so after an extended period, the adrenaline and insulin levels will evidently mismatch, resulting in a stable insulin level but a low adrenaline level that causes a sugar overshoot within a specific cell or cells.

The cell will begin fermentation to discard the excess amount. This very cell will become a tumor, since tumors have a high sugar content.

Severe traumatic situations and experiences may appear in a recurring manner, frequently after years, a decade, or decades. To prevent or fight cancer, the patient needs to actually rest and be happy to raise their adrenaline level. And the primary contributor to this is *money*. Every individual is pressured to enter a specific area with unsatisfactory well-being, which also contributes to depression. Low average salaries and low recognition in education contributed to higher suicide rates. In contrast, relatively stable ones stabilize the suicidal individual or average individual, but put more strain on their mental health and result in stress once more. But there is more to it.

The main reason this has never been discussed is that money, in general, carries power and recognition and is used for one's own intended purposes. But what purpose does money serve when it is limited in access and only leads to more restrictions? It only limits our well-being, even that of powerful and influential individuals, and in the end, contributes to suffering, global warming, and no further evolutionary progress in us humans.

We cannot evolve and improve if we are restricted by it. A potential future outcome may involve humans evolving *backwards*, while the environment is adapted to the previous evolutionary state. This would be a big crash. What advantages would this concept have besides a healthier world? Not only fair competition but also true passion, chaperoned to different individuals, peers, and the whole world. Climate change and significant innovations to stop and arguably reverse it would emerge. Science could finally prove various theories to accelerate efficient processes across numerous fields, with applications in everyday life and objects.

This concept is broad and would act as a filter for passion, dedication, goodwill, and, thereby, safety, preventing a lack of well-being and uncontrolled scenarios. World hunger would be easier to manage and neutralize. Education deserves the attention it needs, and new concepts that are fair and highly interesting, that incorporate the idea of creating a foundation for everything, are no longer a dream. Competition would be controlled and actually contribute to the collective good. War is practically nonexistent. The crime rate and theft would decrease, reducing pressure on individuals.

This concept is dependence on independence, collectively depending on contribution which can also be independently akin to a fixed state. And complications with this concept can be resolved seamlessly because of its fundamental freedom. One might ask how this could work out for individuals who lack interest in pursuing something. That is affirmative and totally passable.

This is where the concept of "discovery centers" comes into play. These hypothetical discovery centers would not be filled with stations; instead, they would be open and designed to feel intriguing to *anyone*, regardless of age or country of origin. Ideally, they are placed in free regions adjacent to large cities.

Why do any task if everything is free? It has to be rephrased this way; the possibility of depression and other illnesses would increase in any individual, and by contributing something and receiving something in return, this can foster a greater feeling towards everything and provide meaning due to the dependence on independence, in which everything can be enjoyed.
Discovery centers convey possibilities and options, or encourage entering into something unique, such as artistic endeavors, that is not generic or common. These centers also have support areas, along with therapy, to

truly support all individuals who need it. The discovery center can also foster social exchange and thus friendship, and can also collaborate with educational institutions to provide worry-free inspiration for children and adults. Anyone will get the chance to become whatever they want to contribute to, inherit, and get something in return —a contribution consisting of many contributions, like an apple, for instance. It is of importance, because the most significant part of contributing to society and the world is recurrently layered with the objective of "must do"; a thought that degrades the world over time and poses risks, as mentioned earlier. The other reason is that not all individuals have the financial means to afford the insight and education they need, which notably increases the sector rate (aggregate cost pressures) for *significant* infrastructure projects.

Moreover, this also increases honesty, health, opportunities, advancements, and accelerates collective evolution. Traditions, information, and a coherent exchange for our future. There is no room for money, power, and egoism, except for living in loneliness and holding on to financial resources as the primary means of expression. It leads to the extinction of nearly anything we might cherish and consider important.

Additionally, one might ask how this could ever work, given its unlikelihood. First of all, it appears to be a step toward solving the majority of problems that hinder anything from adjusting accordingly to the future and human needs. The strategy for executing this is undoubtedly complex but potentially worthwhile. Additional laws and rules must be enforced and removed, though not everything has to be changed. This procedure could start with a city or country to demonstrate its potential, and it is implemented incrementally to ensure a smooth transition. A priority categorization would allow us to identify which sectors can adopt this money-free concept without leaving a crater or incinerating the entire infrastructure, so that, in the end, all sectors experience a smooth transition. Of course, votes need to be taken into consideration before beginning these procedures.

Essential utilities, such as water and electricity usage, would be adjusted accordingly based on the number of household members, health status, and age, with a high limit or margin.

If that generous margin is surpassed, an inspection for water leakage or a damaged electricity meter would be performed after surveying and interviewing the members. If it should turn out to be the members' fault, a warning will be issued. Suppose such a thing should happen within the next quarter of the year. In that case, the usage limit will be reduced to the average usage for the same categories, such as four household members: 1 below 18 years of age, 2 above 18 years of age, and 1 above 60 years of age, with member number 3 physically incapacitated. It also accounts for family visits and more. Indeed, further exploration of categories and studies could make sharing and usage simpler.

This potential system can also be applied to household products of any kind, while accounting for the living area's square meters. We have these numbers and no further use for implementing them in everyday life autonomously to reduce parallel thinking. In contrast, one might have more important tasks to fulfill. These acquisitions will be handled via a citizen ID card that also serves as a biometric verification to ensure safe and fair acquisitions. These identification cards are linked to family and household information and display a curious individual the current state, amount, and acquisitions. Other acquisitions, such as software and digital services, are unique and can be acquired by any individual once.

Categorizing further, there are essential goods and unessential ones. A phone is essential, whereas a private laptop is unessential, as companies can provide them to employees. This applies to the banality of goods. Rare jewelry is unessential, though a vehicle or vehicles, adjusted to household and availability, is essential, as well as suitable insurance and a driver's license.

As rarity increases and availability decreases, a new strategy must be introduced to arguably preserve fairness. Lottery is a great concept, but its application leads to chaos. So, what if applicants, locally or online, enter for certain products and goods, which are then anonymously pooled to determine the final candidates? Furthermore, an age restriction would be derived from the citizen ID card and apply to items, objects, or any medium, whether on a platform or via pooling.

What about used items, etc.? They can still be offered on various platforms as is or in exchange for another item, object, medium, or otherwise. And what definitive reason gives the possibility that this concept is feasible, even in its bare form? It is the dependence of independence. Every individual has the chance to contribute something physically, verbally, digitally, all of the mentioned and more. It is a new form of causal chain in which all aspects equalize and, conceivably, create a foundation for almost anything. On the verge of future trajectories, there are several more possibilities to collectively fill uncertain gaps. The most significant aspect is to passionately and calmly enforce measures that improve the quality of life worldwide and increase the well-being rate as we continue to evolve.

Conclusion and a Foundation for Big Questions

The theories regarding the Universe and thus the foundation of everything within, which has here been set forth, is only advanced as a mere speculation, as a possible hypothesis.

Even when there is insecurity, finding a palpable answer for everything that makes up reality is relatively complicated. We are left alone in this vast space of many things with no other species. We feel alone, we make up theories to believe in something that fundamentally alters us in every way.

Finding purpose in life is a key so unique and broad that there is no coherent definition of what we are or what consciousness is, and it is, more or less, something that needs to be defined by the one asking these questions.

We never chose to be here, yet we are. Due to so many different stimuli and patterns, there is always something that can differ from the remaining and standardized ones.

Even when we die, energy is never lost, and being deceased can still change uncertain parallels of reality and us, because of the potential influence of energy in all kinds of instances of the perceivable reality.
Finding joy and being joyful are the most essential foundations of living a healthy life full of intense, overwhelming, and unique moments.

Even if we are all different, we are and were all born on this very planet and have more in common than some might assume.
This mechanical friction of difference is the reason for our evolution and evolvement. Disparity shapes all of the absence banalities we do not recognize at first sight. We were what we were, we are what we are, and we will be what we will be. In the end, everything and anything has a reason and a meaning.

Glossary

A

Accumulate — to gather or build up over time.

Acquisition — the act of gaining or obtaining something.

Ad infinitum — over and over; without limit.

Adjacent — next to or very near.

Adrenaline — a hormone released in stress or excitement, increasing heart rate and energy.

Agony — extreme physical or emotional pain.

Akin — similar in nature or character.

Alter — to change something.

Alterations — changes or adjustments.

Alternating — happening in turns, one after another.

Amplitude — the height or strength of a wave or vibration, represented as a curve.

Atomic — relating to atoms, the smallest units of matter.

B

Banal — ordinary and unoriginal; monotonous.
Biased — showing unfair preference for one side.

C

Calibration — adjusting a tool or system for accuracy.
Cardinal — something of great importance.
Carbohydrates — sugars and starches that provide energy.
Celestial bodies — natural objects in space, like stars or planets.
Cognitive faith — trust in ideas or reasoning without direct proof, as the initial idea serves as proof.
Coherent — logical and consistent.

Collective — relating to a group acting together.
Collectively — as a group; all together.
Compensate — to make up for something or balance it.
Conceptualized — formed into a clear idea.
Concise (laws) — expressed briefly and clearly.
Conspicuous — easy to see or notice.
Constrained — restricted or limited.
Convey — to communicate or carry.
Conviction — a strong belief; also a legal finding of guilt.
Correlation — a mutual relationship or connection.
Curate — selected terms or any process that involves selecting.

D

Dedicate — to commit time or effort to something.
Delusional — strongly believing something false.
Demand — a strong request or need.
Derive — to obtain from a source.

Dilation — the act of expanding or widening.
Divergence — moving apart or becoming different.
Diversity — variety; a range of other things.
Disparate — very different in kind.
Discourse — written or spoken communication or debate.
Dispersed — spread out over an area.
Detrimental — tending to cause harm.
Distinct — clearly different or separate.

E

Entropy — disorder or randomness in a system.
Evidently — clearly; obviously.
Excess — more than is needed.
Explicit — stated clearly and directly.

F

Finite — having limits or an end.
Fragmented — broken into pieces.

G

Glucose — a simple sugar that provides energy.
Glycogen — a stored form of glucose in the body.
Gravitational time dilation — the phenomenon in which time passes more slowly in a stronger gravitational field.

H

Habitable — suitable for living.

Haywire — out of control or not working correctly.

Hinder — to block or slow progress.

I

Idle — not active or in use.

Incinerate — to burn completely.

Incapacitated — unable to act or function.

Infringing — violating rules or rights.

Inbound — coming in or approaching.

Inorganic — not from living matter.

Insulin is a hormone that regulates blood sugar.

Intercept — to stop or catch before arrival.

Intertwined — closely linked or twisted together.

Invasive — spreading aggressively and causing harm.
Inherit — to receive from someone before.

L

Legacy — something handed down from the past.
Loop quantum gravity — a theory describing space and gravity using quantum mechanics.

M

Mandatory — required; not optional.
Margin — the edge or border; also, the difference between amounts.
Mere — nothing more than; only.
Metabolized — processed by the body's chemistry.
Meticulous — very careful and precise.

Mutual — shared by two or more.

O

Occur — to happen.

Organic — from living matter.

Overlap — partly cover or share the same area.

P

Palpable — able to be felt or easily noticed.

Plane — a flat surface or level.

Planck length — the smallest measurable unit of length in physics.

Pooled — gathered together for common use.

Predecessors — those who came before.

Preface — an introduction at the beginning of a book.
Prolong — to make something last longer.
Propellant — a substance that pushes something forward.

Q

Qubits — quantum bits that can exist in multiple states at once.

R

Radii — plural of radius; lines from a circle's center to its edge.

Recurring — happening over and over.

S

Sentience — the ability to feel and perceive.
Singularity — a point where standard rules break down, often in physics.
Stimuli — things that cause a response.
Stimulus — a single thing that causes a reaction.
Subconscious — mental activity below full awareness. Substitute — a replacement.
Successor — one who follows another.
Superposition — a quantum state of being in multiple conditions at once.
Surpass — to go beyond.
Symbiosis — close living together that benefits one or both.
Substantial — large in amount or importance.

T

Trajectory — the path of a moving object.
Transpired — happened or became known.
Three-body problem — the difficulty of predicting the motion of three interacting bodies.

V

Vast — very large.
Veracious — truthful and accurate.
Verge — the edge or border.

W

Wandered — moved without a fixed direction in unpredictable patterns.

About the Author

Pascal Natis has published his very first book, named Total Theories of Things and Life: The Infinite Existence of the Universe and Our Finite Experience

Pascal is a polymath and autodidact, constantly eager to maximize the many parallels, connections, and knowledge without a quota across various areas, and to create his own theories, things, and aspects. His interests span almost every field, and he is on a mission to discover many more and apply them in many ways. Since he could think, he embarked on a journey of understanding anything, while curiosity became a significant part of this. This does not involve a coherent understanding, but rather connecting anything to its origin and parallel partner, which is why it may appear to switch between directions of inheriting and conveying knowledge.

His curiosity has always kept his mind occupied and has entirely removed the term "relaxation" from his vocabulary.

Pascal has never found another individual to exchange ideas with on various topics —not only on facts, but well beyond standard conversations —and this has left him cognitively unstimulated. This void has fueled his desire to serve humanity, to show how vast anything can be, given that any form of suppression limits the majority. The desire to serve later became a mission, this book included, in which he submerged himself in a chaotic labyrinth of anything he could think of to convey briefly, while making sure to have a unique piece of work rather than one of deceased individuals who shared similar traits and expressions, not only support, but also the realization of something and nothing.

At the same time, the concept of superposition is something that can be found anywhere, to give a clearer understanding of things that were as unclear as they didn't exist.

Live great and fulfilled, experience, think, unravel the hidden, and dream.

www.ingramcontent.com/pod-product-compliance
Lightning Source LLC
Chambersburg PA
CBHW031329230926
51848CB00001B/162
9783000850004